S0-CXX-662

INVESTIGATING EARTH

Air

Kate Walker

This edition first published in 2012 in the United States of America by Marshall Cavendish Benchmark
An imprint of Marshall Cavendish Corporation
All rights reserved.

No part of this publication may be reproduced, stored in a retrieval system or transmitted, in any form or by any means, electronic, mechanical, photocopying, recording, or otherwise, without the prior permission of the copyright owner. Request for permission should be addressed to the Publisher, Marshall Cavendish Corporation, 99 White Plains Road, Tarrytown, NY 10591. Tel: (914) 332-8888, fax: (914) 332-1888.

Website: www.marshallcavendish.us

This publication represents the opinions and views of the author based on Kate Walker's personal experience, knowledge, and research. The information in this book serves as a general guide only. The author and publisher have used their best efforts in preparing this book and disclaim liability rising directly and indirectly from the use and application of this book.

Other Marshall Cavendish Offices: Marshall Cavendish International (Asia) Private Limited, 1 New Industrial Road, Singapore 536196 • Marshall Cavendish International (Thailand) Co Ltd. 253 Asoke, 12th Flr, Sukhumvit 21 Road, Klongtoey Nua, Wattana, Bangkok 10110, Thailand • Marshall Cavendish (Malaysia) Sdn Bhd, Times Subang, Lot 46, Subang Hi-Tech Industrial Park, Batu Tiga, 40000 Shah Alam, Selangor Darul Ehsan, Malaysia

Marshall Cavendish is a trademark of Times Publishing Limited

All websites were available and accurate when this book was sent to press.

Library of Congress Cataloging-in-Publication Data

Walker, Kate.
 Air / Kate Walker.
  p. cm. — (Investigating Earth)
 Includes index.
 Summary: "Describes what air is and why it's important"—Provided by publisher.
 ISBN 978-1-60870-558-0
 1. Air—Juvenile literature. I. Title.
 QC161.2.W34 2012
 551.5—dc22
                                           2010044216

First published in 2011 by
MACMILLAN EDUCATION AUSTRALIA PTY LTD
15–19 Claremont Street, South Yarra 3141

Visit our website at www.macmillan.com.au or go directly to www.macmillanlibrary.com.au

Associated companies and representatives throughout the world.

Copyright text © Kate Walker 2011

Publisher: Carmel Heron
Commissioning Editor: Niki Horin
Managing Editor: Vanessa Lanaway
Editor: Helena Newton
Proofreader: Kylie Cockle

Designer: Kerri Wilson
Page layout: Romy Pearse
Photo researcher: Legend Images
Illustrator: Andrew Hopgood
Production Controller: Vanessa Johnson

Printed in China

Acknowledgments
The author and publisher are grateful to the following for permission to reproduce copyright material:

Front cover photograph: Girl blowing dandelions © Shutterstock/BestPhoto1.

Photographs courtesy of: Corbis/Frank and Helena/cultura, 6; Dreamstime.com/Chiyacat, 4 (center), 18 (right above), /Idrutu, 21, /Janine99, 22, /Klotz, 4 (center right), 18 (right below), /Km2008, 4 (bottom left), 18 (bottom), /Rozaliya, 4 (bottom right), 18 (left above), /Solarwindstudios, 30, /Tommason, 19 (center left), /Voyagerix, 28; Getty Images/Pete Turner, 26; iStockphoto/Elena Elisseeva, 3, 19 (top), 20, /shannon forehand, 9, /Shaun Lowe, 19 (bottom right), /orix3, 5, /Yarinca, 4 (top); Photolibrary/David Deas, 29, /Imagesource, 24, 27, /Lineair, 19 (bottom left), 23; Shutterstock/BestPhoto1, 1, /Ali Ender Birer, 18 (center), 19 (center right), /Susan McKenzie, 4 (bottom center), 18 (left below), /Dudarev Mikhail, 25, /Galushko Sergey, 4 (center right), 18 (top), /Maria Skaldina, 12, /Brykaylo Yuriy, 14.

While every care has been taken to trace and acknowledge copyright, the publisher tenders their apologies for any accidental infringement where copyright has proved untraceable. They would be pleased to come to a suitable arrangement with the rightful owner in each case.

135642

Contents

Investigating Earth	4
Air	5
What Is Air?	6
What Is Air Made Of?	8
How Air Moves	10
How Warm Air Rises	12
How Cool Air Sinks	14
Air Masses and Wind	16
Why Is Air Important?	18
Air Helps Plants Grow	20
Air Helps Animals and Humans Survive	22
Air Affects Weather	24
Air Protects Earth from Space	26
Protecting Air	28
Amazing Air	30
Glossary	31
Index	32

When a word is printed in **bold**, you can look up its meaning in the Glossary on page 31.

Investigating Earth

We investigate Earth to find out what makes it work. Earth is made from natural features. Some of these natural features are living and some are nonliving things.

By investigating Earth we can learn about how it works.

Air

Air is a natural feature of Earth. The whole Earth is surrounded by a blanket of air. Air fills every open space on Earth.

Air surrounds Earth like a huge, clear blanket.

What Is Air?

Air is a nonliving thing. It is a mixture of invisible **gases**. We cannot see air but we can feel air and we can see things being moved around by it.

On windy days we can feel air push against us and see leaves being blown around.

The blanket of air that surrounds Earth is called Earth's **atmosphere**. It is made up of four different layers.

Earth's atmosphere is around 311 miles (500 kilometers) thick.

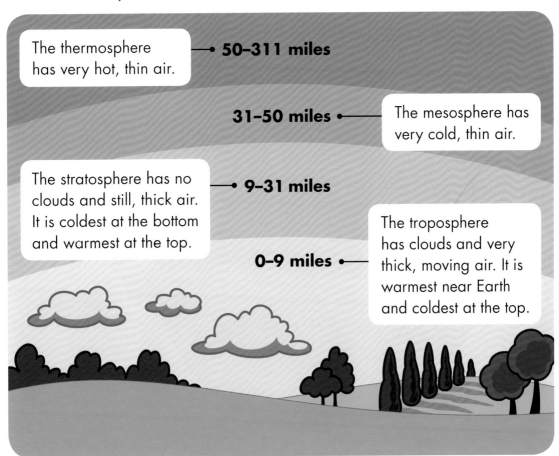

The thermosphere has very hot, thin air. → 50–311 miles

31–50 miles ← The mesosphere has very cold, thin air.

The stratosphere has no clouds and still, thick air. It is coldest at the bottom and warmest at the top. → 9–31 miles

0–9 miles ← The troposphere has clouds and very thick, moving air. It is warmest near Earth and coldest at the top.

What Is Air Made Of?

Air is mostly made from two gases, **nitrogen** and **oxygen**. Air also contains small amounts of other gases. These include **carbon dioxide** and **water vapor**.

There are different amounts of each gas in the air.

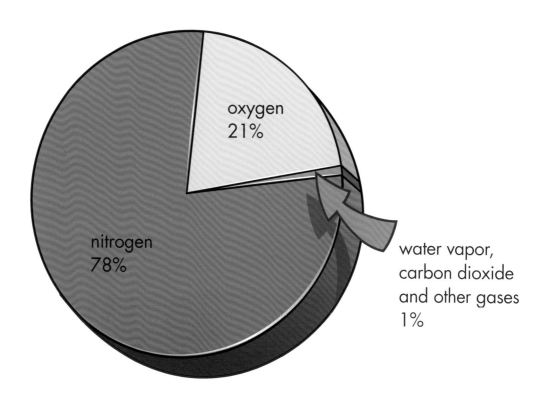

The gases that make up air have no taste or smell. This is why air itself has no smell. However, air can carry smells, such as the scent of pine trees.

Air can carry the scent of pine trees across a meadow.

How Air Moves

Air is always moving around Earth. As air travels around Earth it moves in two main ways.

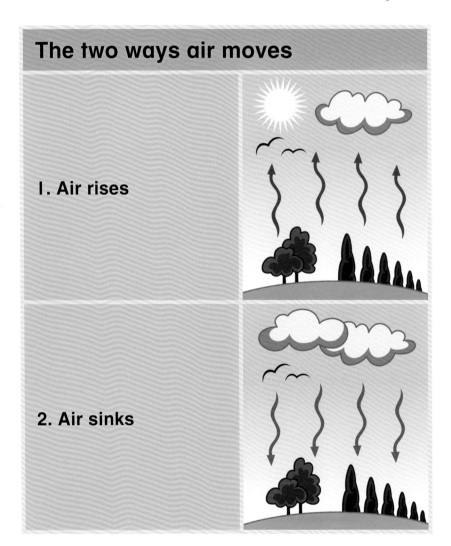

The two ways air moves

1. Air rises

2. Air sinks

Air can have a warm or cool **temperature**. Air rises or sinks because of its warm or cool temperature.

1. Warm air rises.

2. Cool air sinks.

How Warm Air Rises

Air is made warm by sunlight. Sunlight warms Earth and Earth warms the air above it.

In hot, sunny places, such as tropical islands, Earth makes the air above it warm.

When air is warmed it becomes thinner. This makes it lighter than cool air. Warm, light air rises away from Earth.

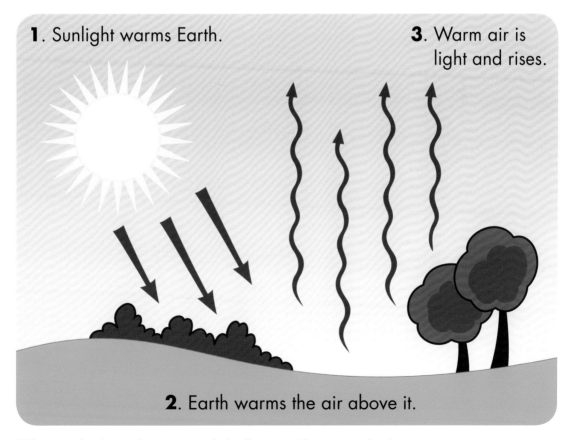

1. Sunlight warms Earth.
3. Warm air is light and rises.
2. Earth warms the air above it.

Warm air rises because it is lighter than cool air.

How Cool Air Sinks

Air cools when there is less sunlight. Less sunlight makes Earth cold and Earth cools the air above it.

In very cold places, Earth makes the air above it cool.

When air is cooled it thickens. This makes it heavier than hot air. Cool, heavy air sinks toward Earth and rests against it.

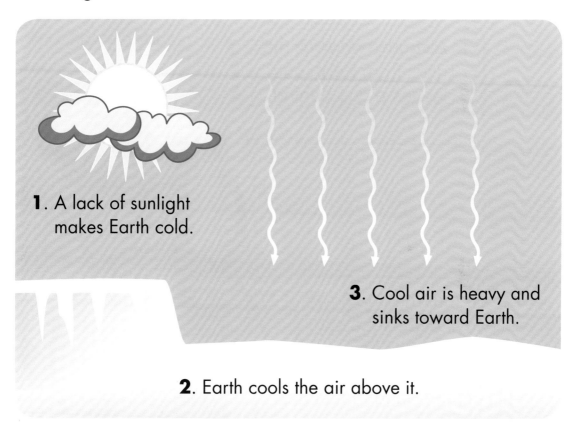

1. A lack of sunlight makes Earth cold.

2. Earth cools the air above it.

3. Cool air is heavy and sinks toward Earth.

Cool air sinks because it is heavy.

Air Masses and Wind

Air masses are giant pools of air that form above Earth. Wind is created by air moving between warm air masses and cool air masses.

1. Air from a warm air mass rises from Earth and leaves a gap behind.

4. The air that rose up flows sideways across the sky. It does this until it can join a mass of cool, sinking air. This creates wind that blows high above Earth.

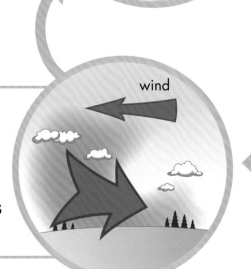

Wind moves across Earth's surface and also across the sky, high above Earth.

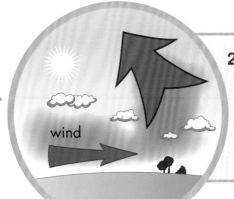

2. Air from a nearby cool air mass flows across the ground to fill the gap. This creates wind that blows across Earth's surface.

3. The rising air gets cool and heavy as it leaves warm Earth behind. It stops rising but cannot sink because more warm air is rising behind it.

Why Is Air Important?

Air surrounds the whole Earth like a blanket of invisible gases. Air works together with some of Earth's other natural features. This helps to keep Earth healthy.

Earth has six main natural features that work together to keep Earth healthy.

Air works together with plants and weather. It also helps animals and humans survive, and protects Earth.

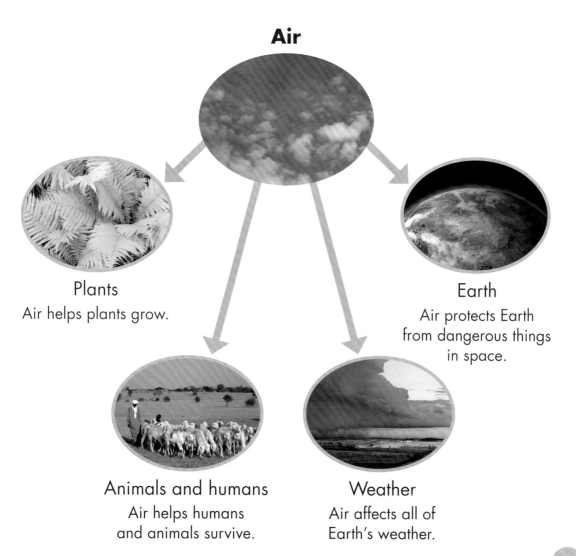

Air

Plants
Air helps plants grow.

Earth
Air protects Earth from dangerous things in space.

Animals and humans
Air helps humans and animals survive.

Weather
Air affects all of Earth's weather.

Air Helps Plants Grow

Plants need **carbon** to grow. Air contains carbon in the form of carbon dioxide gas. Plants get carbon from the air by taking in carbon dioxide. They use carbon to make food.

Trees grow tall by making food from carbon they take from the air.

Plants need sunlight to survive. Air helps spread sunlight by bouncing the rays of light around. This means that light can reach plants in lots of different places.

Air spreads sunlight into shaded places so more plants can grow.

Air Helps Animals and Humans Survive

Animals and humans need oxygen to survive. Air contains oxygen. Animals and humans breathe in oxygen from the air.

Animals and humans need to breathe oxygen all the time.

Air helps keep the temperature on Earth even. The air in the atmosphere stops sunlight from making Earth too hot. This helps animals and humans survive in sunny places.

Without air, many parts of Earth, such as deserts, would be too hot for animals and humans.

Air Affects Weather

Air can be hot or cold. Hot and cold air make hot and cold weather. Weather changes as different masses of hot and cold air move around Earth.

TV weather presenters use weather maps to explain how air movements will affect weather.

Air creates wet weather. Air carries water in the form of water vapor. Clouds form when water vapor changes into water droplets or ice pieces. Clouds then drop water as rain, snow, and hail.

Air and water work together to make cloudy and wet weather.

Air Protects Earth from Space

Air protects Earth from falling **meteors**. Meteors fall from space into Earth's atmosphere every day. The gas in the mesosphere causes speeding meteors to break up before reaching Earth.

Meteors breaking up in the mesosphere are sometimes called shooting stars.

Some of the sun's rays can damage living things. Air in the stratosphere protects Earth from these harmful rays. It stops these rays from reaching Earth.

The air of the stratosphere protects people and plants from the sun's harmful rays.

Protecting Air

Air must be protected because all living things on Earth need clean air. Humans put harmful substances into the air from factories and power plants. These substances cause **air pollution**.

Air pollution, such as smoke from factories, can damage plants and make people ill.

Many harmful substances are used to make new products. If we recycle used products, fewer new products are made. This means fewer harmful substances are released into the air.

We can help protect the air by recycling used products.

Amazing Air

Auroras are moving streams of colored lights that appear in the air. Auroras usually happen above the north and south poles. Auroras are caused by electrical storms that come from outer space.

Auroras happen more than 62 miles (100 kilometers) above Earth.

Glossary

air masses — Large masses of warm or cool air.

air pollution — Substances in air that harm living things.

atmosphere — Air that surrounds Earth.

auroras — Moving colored lights in the air, usually above the north and south poles.

carbon — Substance found in nature and in all living things.

carbon dioxide — Gas in the air that plants take in.

gases — Light, floating substances, such as oxygen or carbon dioxide.

meteors — Lumps of rock or metal from space that have entered Earth's atmosphere.

nitrogen — Gas in the air.

oxygen — Gas in the air that animals and humans breathe in.

temperature — How warm or cool something is.

water vapor — Water that has changed into an invisible gas in air.

Index

A
air masses, 16–17, 24
air movements, 10–14, 16–17, 24
air pollution, 28–29
animals, 19, 22–23
atmosphere, 7, 23, 26
auroras, 30

C
carbon, 20
carbon dioxide, 8, 20
clouds, 7, 25

G
gases, 6, 8–9, 18, 20

H
hail, 25
humans, 19, 22–23, 28

L
living things, 4, 27, 28

M
mesosphere, 7, 26
meteors, 26

N
natural features, 4–5, 18–19
nitrogen, 8
nonliving things, 4, 6

O
oxygen, 8, 22

P
plants, 18–19, 20–21, 27, 28

R
rain, 25
recycling, 29

S
snow, 25
space, 19, 26, 30
stratosphere, 7, 27
sunlight, 12–15, 21, 23, 27

T
temperature, 11, 23
thermosphere, 7
troposphere, 7

W
water, 18, 25
water vapor, 8, 25
weather, 18–19, 24–25
wind, 6, 16–17

ML   12-14